THIS BOOK BELONGS TO

_____

_____

WELCOME

First of all, I would like to thank you for buying this book. You are about to embark on a fantastic coloring journey through our oceans.

COLOR YOURSELF HAPPY

Grab your pencils and leave everyday life behind you.
There's no wrong way to color, just have fun.

PAPER CHOICE

The paper selection at Amazon is very limited. Therefore this book is printed on standard paper. To prevent the color from bleeding through to the image underneath, you should place a few blank sheets of paper under the image you are currently working on.

SHARE YOUR ARTWORK

I'm really looking forward to seeing your colored artwork on social media.
Use the hashtag **#GabiWolfSeaLife** so that I can find your pictures.

CONTACT

Please feel free to contact me if you have any questions or suggestions.

www.gabiwolf.de

@gabisgrafiken

SOAP

Color Test Page

Color Test Page

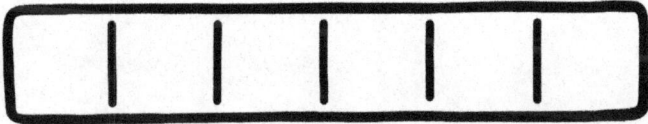

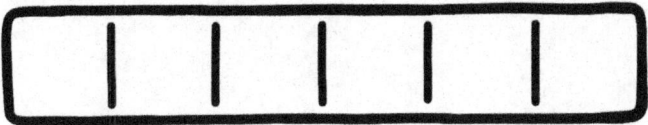

Color Test Page

Please Leave Your Amazon Rating

I am very happy about an honest Rating on Amazon.
Only through feedback can I respond to wishes and suggestions.

Thank you so much for your support!

We have now reached the end of the book

I hope you enjoyed the illustrations and had a lot of fun coloring them. Pleas have a look at my other coloring books. You can find information about them on the following pages, on my website, on Amazon or on social media.

www.gabiwolf.de

@gabisgrafiken

More Coloring Books from Gabi Wolf

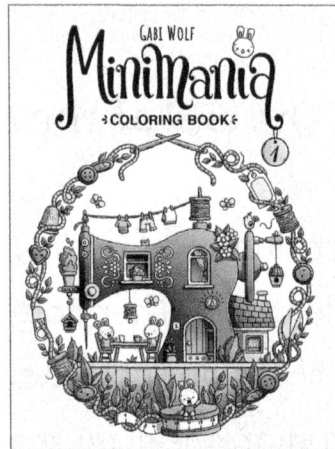

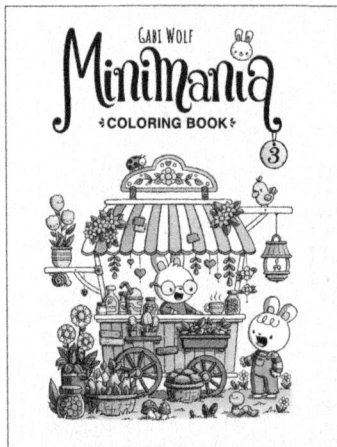

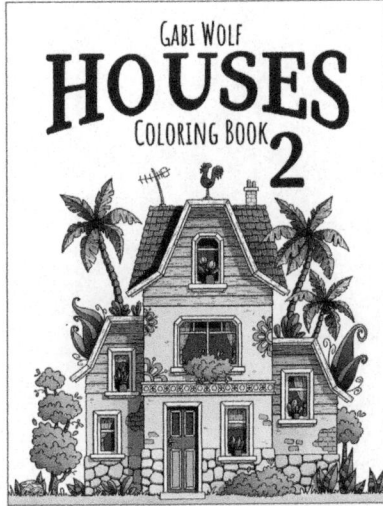

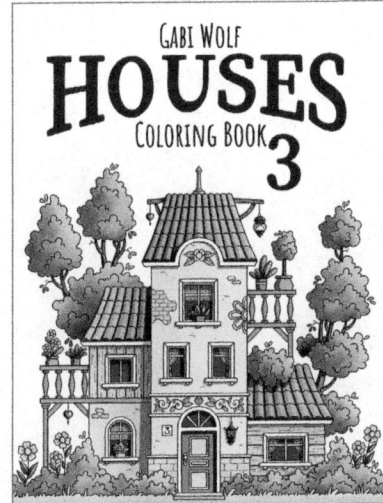

GABI WOLF Volume 1
Wonder Worlds
Coloring Book

GABI WOLF Volume 2
Wonder Worlds
Coloring Book

GABI WOLF
HOUSES
Coloring Book 1

GABI WOLF
HOUSES
Coloring Book 2

GABI WOLF
HOUSES
Coloring Book 3

Illustrations by
Gabi Wolf
Daydreams
Coloring Book

Mandalas
Ein Ausmalbuch für Erwachsene
Band 1

Mandalas
Ein Ausmalbuch für Erwachsene
Band 2

Gabi Wolf

Gabi Wolf

Made in the USA
Coppell, TX
24 December 2024